INTELLIGENT SYSTEMS PRESS

— An Intelligent Systems Press Requirements Discovery Guide —

PROJECT REQUIREMENTS ANALYSIS BOOK OF QUESTIONS

Asking Better Questions, Defining Better Requirements, and Building Better Processes

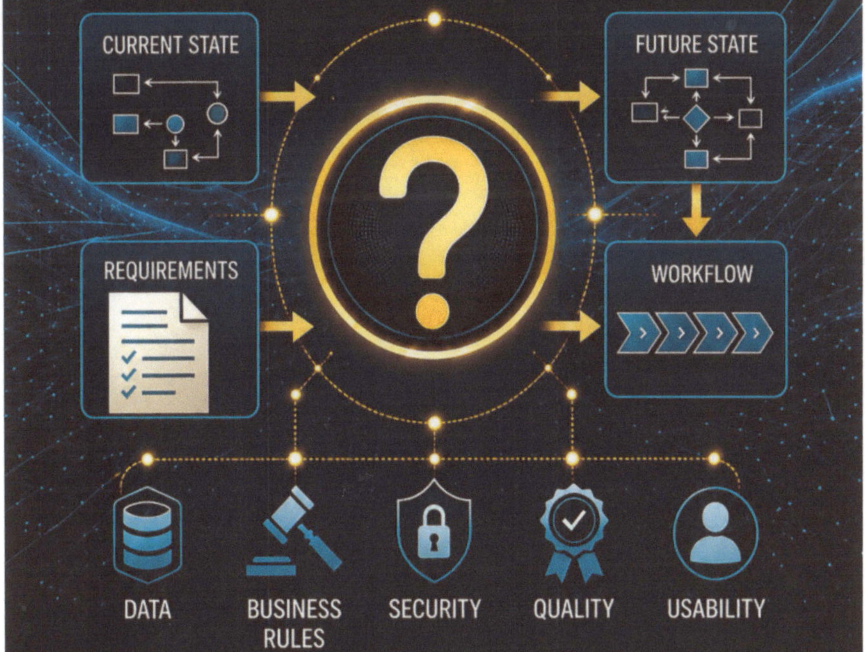

V. QUAN LEE & J. KAY

PROJECT REQUIREMENTS ANALYSIS BOOK OF QUESTIONS

Asking Better Questions, Defining Better Requirements, and Building Better Processes

An Intelligent Systems Press Requirements Discovery Guide

Written by V. Quan Lee and J. Kay

Restored and Expanded Intelligent Systems Press Edition
Intelligent Systems Press

Copyright Page

Restored and expanded edition prepared for Intelligent Systems Press.

Published by:

Intelligent Systems Press

Process Engineering • Requirements Analysis • Intelligent Process Improvement • Process Intelligence

ISBN: 979-8-9958725-7-3

Editing by: Brian Loveless

Web Content Design by: Erik Victor Jones

Content Editing and Layout by: Emma Victoria Lee

DEDICATION

This book is dedicated to every business analyst, process engineer, project manager, product owner, quality professional, subject matter expert, sponsor, developer, tester, and executive leader who has ever had to pause a project and ask:

"Do we truly understand what we are building, why we are building it, who it serves, what problem it solves, and what will happen if we get it wrong?"

Questions are not a delay in the work.

Questions are the beginning of the work.

AUTHOR'S NOTE TO THE RESTORED INTELLIGENT SYSTEMS PRESS EDITION

The original purpose of *Project Requirements Analysis Book of Questions* was simple but important: to help project teams ask better questions before they attempted to define requirements, design systems, improve processes, or deliver solutions.

That purpose remains unchanged.

What has changed is the world surrounding the requirements process. Organizations now operate in environments shaped by automation, Agile delivery, DevOps, artificial intelligence, cloud systems, data platforms, cybersecurity concerns, regulatory pressure, customer expectations, and executive demand for faster delivery. Yet despite all of these changes, one truth remains constant:

Projects still fail when teams do not understand the problem.

Requirements still fail when teams do not ask the right questions.

Technology still fails when the process, people, data, rules, risks, and future state are not understood before the solution is built.

This restored edition honors the original book's question-centered foundation while positioning it within the broader Intelligent Systems Press process engineering series. It is not intended to replace formal requirements methods, business analysis frameworks, Agile practices, software delivery models, or enterprise governance systems. Instead, it strengthens them by returning to the foundation beneath all of them: disciplined inquiry.

Before an organization can improve a process, automate a workflow, modernize a system, align teams, govern delivery, or apply Process Intelligence, it must first learn how to ask the right questions.

That is the purpose of this book.

Table of Contents

HOW THIS BOOK FITS THE INTELLIGENT SYSTEMS PRESS PROCESS ENGINEERING SERIES

The Intelligent Systems Press process engineering series is built around a simple but powerful idea:

Organizations cannot intelligently improve what they have not clearly understood.

Each book in the series addresses a different layer of that understanding.

Project Requirements Analysis Book of Questions is the requirements discovery foundation. It helps teams ask the right questions before requirements are written, designs are approved, systems are developed, or processes are changed.

Feng Shui for Intelligent Process Improvement explores the balance, discipline, and structure needed to create healthier process environments. It reminds organizations that work has flow, friction, energy, and imbalance.

12 Steps to Intelligent Process Improvement provides a structured improvement method for defining, correcting, measuring, and maintaining process discipline.

Process Intelligence elevates the conversation to the enterprise level, showing how Agile, DevOps, AI, governance, operations, and executive strategy must be aligned through an intelligent operating layer.

Process Intelligence Playbook converts the framework into practical stakeholder guidance for engineers, product owners, project managers, Scrum Masters, QA teams, process leaders, and executives.

Within that series, this book is the front door.

It begins where all improvement begins: with the question.

SERIES POSITIONING STATEMENT

Requirements do not fail because people stop working. They fail because teams begin building before they fully understand the problem, the process, the people, the data, the rules, the risks, the exceptions, the constraints, and the future state they are trying to create.

Project Requirements Analysis Book of Questions restores the discipline of requirements discovery by giving business analysts, project managers, product owners, QA professionals, process engineers, subject matter experts, and executive sponsors a structured way to ask better questions before decisions become design, design becomes development, and development becomes operational reality.

HOW TO USE THIS BOOK

This book is designed as a Requirements Discovery Guide for Process Intelligence, Structured Analysis, and Intelligent Process Improvement.

It can be used before a project begins, during discovery, throughout requirements elicitation, during process improvement analysis, in Agile refinement sessions, during stakeholder interviews, while developing use cases, while preparing user stories, during quality reviews, or as part of project recovery when requirements have become unclear.

You do not need to read it from beginning to end in one sitting.

Instead, use it according to the situation:

- Use the As-Is questions when you need to understand the current state.
- Use the To-Be questions when you need to define the future state.
- Use the Scoping questions when boundaries are unclear.
- Use the Business Driver questions when value, priority, funding, or executive purpose needs clarification.
- Use the User Requirements questions when features, functions, roles, and interactions need to be defined.
- Use the Exception questions when alternate paths, error conditions, overrides, and unusual scenarios must be discovered.
- Use the Data questions when information, interfaces, reports, conversions, warehouses, or systems of record are involved.
- Use the Workflow questions when manual procedures, automated processes, handoffs, and dependencies must be understood.
- Use the Business Rules questions when decisions, constraints, calculations, validations, or policies must be documented.

- Use the Stakeholder questions when people, systems, teams, vendors, regulators, or external groups must be identified.
- Use the User Interface questions when screens, reports, navigation, labels, layouts, and field behavior must be specified.
- Use the Usability questions when ease of use, accessibility, efficiency, and user adoption matter.
- Use the Quality and Performance questions when reliability, scalability, availability, volume, response time, and operational expectations must be defined.
- Use the Security questions when access, audit, privacy, compliance, roles, and data protection must be addressed.
- Use the Technical Solution questions when technology constraints, architecture, platforms, tools, integrations, and environments must be understood.
- Use the Gap Analysis questions when the organization needs to compare where it is today against where it wants to go.
- The purpose is not to ask every question in every project.
- The purpose is to know which questions matter before the team commits to a solution.

Intelligent Requirements Discovery Lifecycle™

From Discovery to Operational Insight

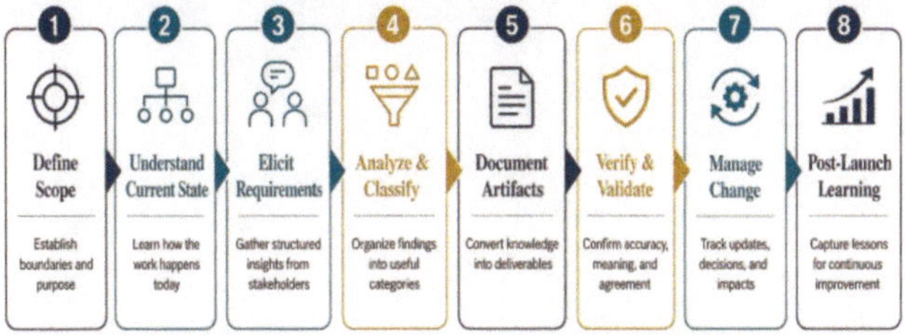

• A lifecycle view of disciplined requirements discovery. •

PART I

THE DISCIPLINE OF ASKING BETTER QUESTIONS

CHAPTER 1

Questions Are the Foundation of Quality Requirements

Every project begins with an assumption.

Someone assumes they understand the problem.

Someone assumes they know what the business wants.

Someone assumes the users agree.

Someone assumes the current process is understood.

Someone assumes the future state is obvious.

Someone assumes technology will solve the issue.

Someone assumes the requirements are complete.

Many projects fail because those assumptions are never challenged.

A requirement is not simply a statement written in a document. A requirement is the result of inquiry, investigation, negotiation, clarification, validation, and agreement. It represents what the organization believes must be true for a solution to deliver value.

The quality of a requirement depends on the quality of the questions that produced it.

Weak questions produce vague requirements.

Vague requirements produce unstable designs.

Unstable designs produce rework.

Rework produces cost, delay, frustration, and loss of confidence.

Strong questions produce clarity.

Clarity produces better requirements.

Better requirements produce better design.

Better design produces better delivery.

Better delivery produces better business outcomes.

This is why questions are not administrative. They are strategic.

CHAPTER 2

The Requirements Analyst as Translator, Investigator, and Process Interpreter

The requirements analyst performs one of the most important roles in any project environment.

The analyst must understand the language of the business and the language of technology.

The analyst must listen to executives, managers, users, subject matter experts, developers, testers, architects, security specialists, operations teams, vendors, customers, and regulators.

The analyst must understand not only what people say, but what they mean.

The analyst must identify contradictions, assumptions, missing details, hidden dependencies, process gaps, unclear ownership, data issues, business rules, risks, and exceptions.

The analyst is part translator, part investigator, part negotiator, part process interpreter, and part quality advocate.

A strong analyst does not simply ask, "What do you want?"

A strong analyst asks:

- What problem are we solving?
- Why does this problem matter?
- Who is affected?
- How is the work done today?
- What happens when the normal process fails?
- What rules govern the decision?
- What data is needed?
- Who owns the data?
- What must be secured?

- What must be measured?
- What must be approved?
- What must remain unchanged?
- What will happen if we do nothing?

The analyst protects the project from premature certainty.

CHAPTER 3

The Cost of Weak Requirements

Weak requirements create invisible debt.

At first, the project may appear to be moving quickly. Meetings are scheduled. Designs are created. Developers begin coding. Testers prepare scripts. Sponsors receive status reports. Everyone feels productive.

Then the questions begin to surface.

Who approved this?

Why does the workflow skip this step?

Where does the data come from?

Why is this field required for one user but optional for another?

Who can override this rule?

What happens when the customer enters invalid data?

Why does the report not match the existing finance report?

Who owns the interface failure?

What happens after launch?

Why was security not involved earlier?

Why did the business expect something different?

These questions are expensive when they are asked late.

They are much less expensive when they are asked early.

The cost of weak requirements is not limited to software defects. It includes:

Rework

Missed deadlines

Budget overruns

User dissatisfaction

Low adoption

Manual workarounds

Duplicate data entry

Security exposure

Compliance risk

Operational confusion

Executive frustration

Loss of trust between business and technology teams

The goal of requirements analysis is not to eliminate change. Change is natural. The goal is to prevent avoidable confusion by discovering what should have been known before the solution was designed.

CHAPTER 4

Requirements Discovery and Process Intelligence

Process Intelligence begins with understanding.

An organization cannot intelligently automate a process it does not understand. It cannot govern a workflow it has not defined. It cannot improve performance if it has not identified the current baseline. It cannot apply AI responsibly if the underlying data, rules, exceptions, and decisions are unclear.

Requirements discovery is therefore not separate from Process Intelligence.

It is the beginning of Process Intelligence.

When analysts ask questions about current state, future state, business drivers, users, data, rules, exceptions, quality, security, and gaps, they are building the foundation for a more intelligent organization.

Each answer becomes a signal.

Each signal helps the organization understand its processes, systems, constraints, risks, and opportunities.

This is why requirements analysis must be treated as more than documentation.

It is enterprise learning.

CHAPTER 5

The Requirements Intelligence Loop™

The Intelligent Systems Press requirements model can be summarized as the Requirements Intelligence Loop™:

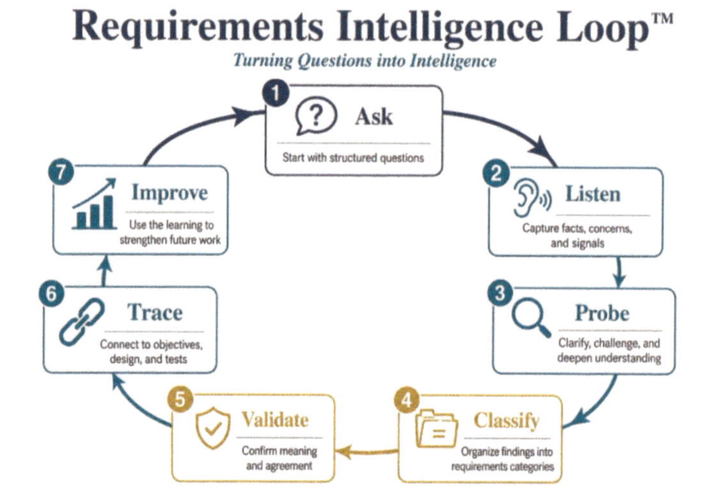

A signature Intelligent Systems Press model for disciplined requirements discovery.

Ask → Listen → Probe → Classify → Validate → Trace → Improve

Ask

Begin with structured questions that reveal the business problem, process reality, user need, data requirement, rule, risk, exception, or constraint.

Listen

Do not treat the stakeholder's answer as a formality. Listen for facts, assumptions, uncertainty, emotion, conflict, dependency, and silence.

Probe

Ask follow-up questions. Clarify vague language. Request examples. Ask what happens before, during, and after the process. Ask what happens when the normal path fails.

Classify

Place answers into requirements categories such as process, data, business rules, security, usability, quality, reporting, workflow, interface, or compliance.

Validate

Confirm understanding with stakeholders. Review the requirement, assumption, rule, or decision before it becomes part of the solution baseline.

Trace

Connect each requirement to its source question, stakeholder, business objective, design element, test case, and acceptance criteria.

Improve

Use what was learned to improve the process, clarify governance, reduce ambiguity, and strengthen future requirements discovery.

The loop is simple, but powerful.

It turns questions into intelligence.

PART II

PREPARING FOR REQUIREMENTS DISCOVERY

CHAPTER 6

Understanding Your Audience

Before asking requirements questions, understand who will use the answers.

Different audiences need different levels of detail.

An executive sponsor may care about strategic alignment, cost, risk, compliance, and business outcomes.

A project manager may care about scope, timeline, dependencies, resources, milestones, and risks.

A product owner may care about user value, prioritization, acceptance criteria, and release planning.

A business user may care about daily workflow, ease of use, process fit, and operational impact.

A developer may care about logic, data structures, interfaces, validations, and technical constraints.

A tester may care about expected results, alternate paths, error handling, boundary conditions, and acceptance criteria.

A security professional may care about access, auditability, privacy, encryption, roles, and compliance.

A process engineer may care about inputs, outputs, handoffs, controls, rework, cycle time, and process variation.

Before writing requirements, ask:

- Who will read this?
- Who will approve this?
- Who will build from this?
- Who will test from this?
- Who will operate this?
- Who will be measured by this?
- Who will be affected if this is wrong?

Requirements should be written for the people who must make decisions from them.

CHAPTER 7

Stakeholder Discovery and RASCI Thinking

Requirements are shaped by stakeholders.

A stakeholder is any person, group, system, organization, or authority that affects or is affected by the solution.

Stakeholders may include:

Executive sponsors

Business owners

Product owners

Project managers

Scrum Masters

Business analysts

Process engineers

Developers

Architects

Database administrators

QA testers

Security teams

Operations teams

Help desk teams

Compliance teams

Legal teams

Vendors

Suppliers

Customers

End users

Regulators

Reporting consumers

Data owners

System owners

Training teams

Change management teams

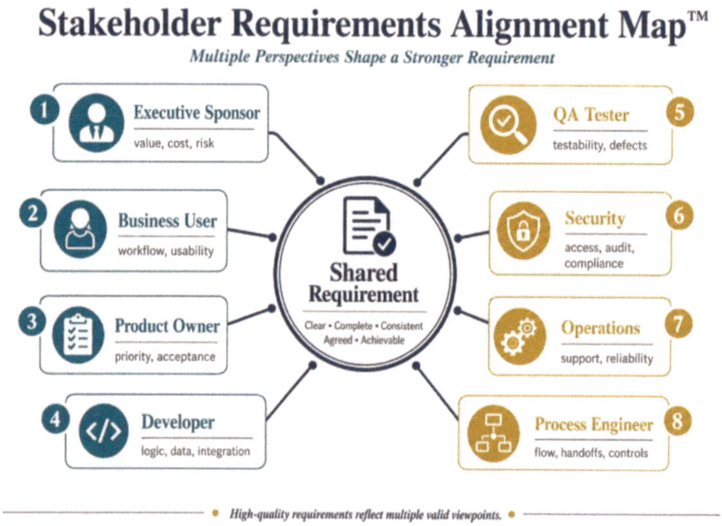

A useful way to organize stakeholder involvement is through RASCI thinking:

- Responsible — Who performs the work?
- Accountable — Who owns the outcome?
- Supportive — Who helps complete the work?
- Consulted — Who provides input before decisions are made?
- Informed — Who must be kept aware of progress, decisions, or impacts?

Requirements problems often occur because the wrong people were consulted, the right people were ignored, or decision authority was unclear.

Before requirements discovery begins, ask:

- Who owns the business problem?
- Who owns the process?
- Who owns the data?
- Who owns the system?
- Who owns the budget?
- Who approves the requirements?
- Who can reject the solution?
- Who can change the priority?
- Who will use the solution daily?
- Who will support the solution after launch?
- Who will be blamed if the solution fails?
- Who has knowledge that is not documented anywhere else?

CHAPTER 8

Planning the Requirements Interview

A requirements interview should not be treated as casual conversation.

It should be prepared, structured, facilitated, documented, and validated.

Before the interview, determine:

The purpose of the interview

The stakeholders who should attend

The decisions that need to be made

The topics to be covered

The questions to be asked

The documents to review

The artifacts to update

The assumptions to validate

The unresolved issues to clarify

The expected outputs

A strong interview agenda may include:

Project purpose

Business problem

Current process

Future-state expectations

Scope boundaries

Stakeholder roles

Data needs

Business rules

Exceptions

Security concerns

Reporting needs

Quality expectations

Risks and assumptions

Open questions

Next steps

- Whenever possible, send questions in advance. This gives stakeholders time to think, gather examples, bring documents, and avoid guessing during the meeting.
- Good requirements interviews are not interrogations.
- They are structured discovery sessions.

CHAPTER 9

Listening, Probing, Clarifying, and Validating

The most important skill in requirements discovery is not speaking.

It is listening.

Many analysts make the mistake of thinking about the next question while the stakeholder is answering the current one. When that happens, important details are missed.

Listen for:

Unclear words

Conflicting statements

Assumptions

Hesitation

Emotion

Repeated complaints

Manual workarounds

Hidden approvals

Exceptions

Unwritten rules

Dependencies

Risk signals

Statements such as "usually," "sometimes," "it depends," "normally," "unless," and "except when"

Those words often reveal requirements.

When a stakeholder says, "The system usually approves it automatically," ask:

- When does it not approve automatically?
- Who reviews it then?
- What rule determines the exception?
- How often does this happen?
- Is it logged?
- Can someone override it?
- Who has authority to override it?
- Does the customer receive a message?
- Does the transaction continue or stop?

Clarifying questions turn vague answers into usable requirements.

CHAPTER 10

Turning Answers into Requirements Artifacts

A requirement should not remain trapped in meeting notes.

Each answer should be converted into the proper artifact.

Some answers become process steps.

Some become business rules.

Some become data elements.

Some become use cases.

Some become user stories.

Some become acceptance criteria.

Some become test cases.

Some become interface requirements.

Some become security controls.

Some become assumptions.

Some become risks.

Some become out-of-scope items.

Some become future enhancements.

Question-to-Artifact Mapping Matrix

How Better Questions Produce Better Deliverables

	Question Category	What It Reveals	Primary Artifact
1	As-Is	current-state reality	workflow diagram
2	To-Be	future-state intent	future-state model
3	Scope	boundaries and assumptions	scope statement
4	Business Drivers	value and priority	business case
5	User Requirements	roles, needs, features	use cases or user stories
6	Exceptions	alternate paths and edge cases	exception log
7	Data	information needs and sources	data dictionary
8	Business Rules	decision logic and constraints	rules catalog
9	Security	access and control	security matrix
10	Gap Analysis	current-to-future gaps	gap analysis worksheet

Map every important question to a usable project artifact.

The analyst must determine where each answer belongs.

A good requirements process does not merely collect information.

It organizes information into a structure that can be reviewed, approved, built, tested, governed, and improved.

PART III

THE CORE BOOK OF QUESTIONS

CHAPTER 11

As-Is Current State Questions

The current state describes how the work is performed today.

Before defining the future, the team must understand the present.

Ask:

- What is the current problem?
- What are the current opportunities?
- What does the product, process, or system currently do well?
- What current features or functions should remain?
- Why should they remain?
- What current features or functions should be removed?
- Why should they be removed?
- How is the work currently being accomplished?
- Who currently performs the work?
- What was the original intent of the legacy process or system?
- What systems are currently used?
- What procedures are currently followed?
- What tools support the work today?
- What reports are currently used?
- What data is currently captured?
- What data is missing?
- What manual workarounds exist?
- Where does rework occur?
- Where do delays occur?
- Where do errors occur?
- Where are users frustrated?
- What controls exist today?
- What approvals exist today?
- What business rules exist today?

- What exceptions exist today?
- What are the current risks?
- Where is the organization vulnerable?
- What happens if the current process remains unchanged?
- What will happen if this project is not completed?

Quick Tip

For each process in scope, document the trigger, objective, actors, inputs, tasks, tools, outputs, business rules, data, quality controls, security requirements, happy path, alternate paths, exception paths, and downstream impacts.

Common Artifacts

- Current-state process map
- Workflow diagram
- Use case description
- Process inventory
- Pain point log
- Legacy system summary
- Current-state data inventory
- Risk and issue log

CHAPTER 12

To-Be Future State Questions

The future state defines what the organization wants the process, system, product, or service to become.

Ask:

- What do we know about the desired future state?
- What problem should the future state solve?
- What opportunity should the future state create?
- What functionality must be included?
- What capabilities are essential?
- What capabilities are optional?
- What must remain the same?
- What must change?
- What should be eliminated?
- What should be automated?
- What should remain manual?
- What should be simplified?
- What should be standardized?
- What resources will be needed?
- What technology will be needed?
- What training will be needed?
- What policies will be affected?
- What procedures will be affected?
- What timeline is expected?
- What is driving the timeline?
- What is at risk if the timeline is missed?
- What are the success factors?
- How will success be measured?
- What is out of scope?
- What is the happy path?
- What are the alternate paths?
- What are the exception paths?
- What data requirements must be included?

- What business rules must be included?
- What security requirements must be included?
- What reporting requirements must be included?
- What operational support will be needed after launch?

Quick Tip

The future state should be specific enough to guide design but flexible enough to allow solution options. Avoid defining a solution before the problem, goal, and requirements are understood.

Common Artifacts

- Future-state process map
- To-Be workflow diagram
- Use case model
- Requirements table
- Traceability matrix
- Solution scope statement
- Future-state operating model

CHAPTER 13

Scoping Questions

Scope defines the boundaries of the project.

Without scope, requirements become a moving target.

Ask:

- What is included in this project?
- What is excluded from this project?
- What are the project risks?
- What are the business impacts?
- What assumptions are being made?
- What constraints exist?
- What dependencies exist?
- What prerequisites must be completed?
- What are the project showstoppers?
- What decisions have already been made?
- What priorities have already been established?
- Who are the decision makers?
- Who is the financial sponsor?
- Who controls the funding?
- What milestones are fixed?
- What deadlines are flexible?
- What deadlines are not flexible?
- What resources are available?
- What resources are missing?
- How much experience does the team have with this type of project?
- Is the technology new?
- Is the business process new?
- Are the requirements stable?
- Are the requirements likely to change?
- How realistic is the solution based on budget, schedule, and resources?

- How will requirements be prioritized?
- How will out-of-scope items be tracked?
- What methodology will be used?
- What tools will be used?
- What commercial products or vendor solutions have already been selected?
- What executive decisions are non-negotiable?
- What target market or user group is affected?
- What environment will the solution operate within?

Quick Tip

Scope should be documented and approved before detailed requirements discovery begins. Without scope agreement, the project team may spend the entire requirements phase chasing a moving target.

Common Artifacts

- Scope statement
- Project charter
- Assumptions log
- Constraints log
- Risk log
- Dependency register
- Out-of-scope register
- Decision log

CHAPTER 14

Goal, Business Driver, and Prioritization Questions

A requirement has more value when it can be traced to a business driver.

Ask:

- What is the goal of this project?
- What do you hope to accomplish?
- What business problem is driving the change?
- What opportunity is driving the change?
- What will happen if the change is not made?
- What is the impact if the change is delayed?
- Who benefits from the project?
- How will success be measured?
- What value will this bring to the organization?
- How does this support enterprise strategy?
- How does this support department goals?
- How does this support customer needs?
- How does this support compliance needs?
- How does this support operational efficiency?
- How does this support revenue?
- How does this reduce cost?
- How does this reduce risk?
- How does this improve service?
- How does this improve quality?
- What is the priority compared with other projects?
- What is the expected budget?
- What financial benefit justifies the cost?
- Is this project driven by regulation?
- Is this project driven by customer demand?
- Is this project driven by competitive pressure?
- Is this project driven by technology modernization?

- If features must be cut, which ones would be removed first?
- Which requirements are showstoppers?
- Which requirements are must-haves?
- Which requirements are should-haves?
- Which requirements are could-haves?
- Which requirements will not be included now?

Quick Tip

Prioritization is not a one-time activity. Priorities may change as risks, costs, dependencies, and stakeholder understanding change.

Common Artifacts

- Business case
- Prioritization matrix
- Value statement
- ROI summary
- Project charter
- Executive briefing
- Requirements ranking table

CHAPTER 15

User Requirements, Features, and Functionality Questions

User requirements describe who needs the solution and what they need to accomplish.

Ask:

- What types of users will use the system?
- Are the users novices, experts, supervisors, managers, administrators, executives, or external customers?
- How will each user group interact with the system?
- What tasks will each user perform?
- What triggers each user action?
- Will triggers come from humans, systems, schedules, events, or interfaces?
- What features are required?
- What features are optional?
- What functions are required?
- What functions should be restricted?
- How will users search for information?
- How will users enter information?
- How will users update information?
- How will users cancel or delete information?
- How will users view reports?
- How will users receive notifications?
- How will users know an action was successful?
- How will users know an error occurred?
- What help will users need?
- What training will users need?
- What user roles are required?
- What permissions are required?
- What functions should be available to each role?
- What functions should not be available to each role?

- What existing system should this resemble?
- What existing system should this avoid resembling?
- How will usability be measured?
- How will user satisfaction be measured?

Quick Tip

Each user requirement should be traceable to a role, business objective, process step, acceptance criteria, and test case.

Common Artifacts

- User stories
- Use cases
- Role matrix
- Functional requirements
- Mockups
- Prototypes
- Storyboards
- Security matrix
- Acceptance criteria

CHAPTER 16

Exception and Alternate Path Questions

Many requirements fail because the team documents only the happy path.

Real systems must handle exceptions.

Ask:

- What are the exceptions to the normal rule?
- What are the alternate paths?
- What are the predictable user errors?
- What errors are commonly reported to the help desk?
- What unusual data may be entered?
- What happens when required data is missing?
- What happens when invalid data is entered?
- What happens when duplicate data is entered?
- What happens when the system is unavailable?
- What happens when an interface fails?
- What happens when a user does not have permission?
- What happens when approval is denied?
- What happens when approval is delayed?
- What happens when a deadline is missed?
- What business rules can be overridden?
- Who can override them?
- Should override authority continue?
- What business rules are likely to change?
- What codes or values are likely to change?
- What exceptions must be logged?
- What exceptions must be audited?
- What transactions must be time stamped?
- What exceptions require management review?
- What exceptions require customer notification?
- What exceptions stop the process?

- What exceptions allow the process to continue?

Quick Tip

The exception path often reveals the real complexity of the process. Do not treat exceptions as minor details.

Common Artifacts

- Alternate path documentation
- Exception log
- Business rules table
- Use case extensions
- Error message catalog
- Audit requirements
- Security requirements
- Test scenarios

CHAPTER 17

Data, Interface, Warehouse, and Conversion Questions

Data is one of the most important parts of requirements analysis.

Ask:

- Where will the data come from?
- Who owns the data?
- Who creates the data?
- Who updates the data?
- Who consumes the data?
- What is the system of record?
- What data must be captured?
- What data is optional?
- What data is mandatory?
- What data is conditionally required?
- What are the data elements?
- What are the characteristics of each data element?
- What are the field lengths?
- What are the data types?
- What values are allowed?
- What values are invalid?
- What are the default values?
- What data must be unique?
- What data may repeat?
- What data must be retained?
- How long must data be retained?
- What data must be archived?
- What data must be purged?
- What data must be converted from another system?
- What data feeds are required?
- What interfaces are required?

- What data flows into the system?
- What data flows out of the system?
- When must data be synchronized?
- How often must data be synchronized?
- What reports use this data?
- What decisions are made from this data?
- What data is needed for analytics?
- What data is needed for compliance?
- What data is needed for AI, automation, or future intelligence capabilities?

Quick Tip

When a stakeholder gives a vague data answer, probe for more detail. Ask for examples, reports, screenshots, forms, spreadsheets, databases, and current system references.

Common Artifacts

- Data dictionary
- Interface inventory
- Data flow diagram
- Source-to-target mapping
- Conversion plan
- Report inventory
- Metadata catalog
- Data ownership matrix

CHAPTER 18

Automated Process, Workflow, and Manual Procedure Questions

A system is rarely just a system.

It is usually part of a larger workflow.

Ask:

- What processes will be automated?
- What processes will remain manual?
- What manual procedures will be updated?
- What manual procedures will become obsolete?
- What new manual procedures will be introduced?
- What are the process timelines?
- What are the process milestones?
- What is the sequence of steps?
- Who performs each step?
- What triggers the process?
- What ends the process?
- What are the inputs?
- What are the outputs?
- What are the dependencies?
- What are the prerequisites?
- What are the decision points?
- What approvals are required?
- What handoffs occur?
- Where does work wait?
- Where does rework occur?
- Where does duplication occur?
- Where does the process cross departments?
- Where does the process cross systems?
- How do manual and automated steps interact?

- How do processes map to use cases?
- How do processes map to data?
- How do processes map to business rules?
- How do processes map to security roles?

Quick Tip

Process improvement and software requirements should be connected. A system that automates a broken process may simply make the broken process fail faster.

Common Artifacts

- Swimlane diagram
- Activity diagram
- Workflow diagram
- Process template
- Data flow diagram
- State transition diagram
- Handoff analysis
- Process inventory

CHAPTER 19

Business Rules and Constraint Questions

Business rules define decision logic, restrictions, calculations, and operating boundaries.

Ask:

- What rules are needed to protect the integrity of the data?
- What rules are needed to protect the integrity of the process?
- What policies govern this process?
- What regulations govern this process?
- What decisions must the system make?
- What decisions must humans make?
- What conditions trigger a decision?
- What formulas are required?
- What calculations are required?
- What validations are required?
- What restrictions apply when adding data?
- What restrictions apply when updating data?
- What restrictions apply when canceling data?
- What restrictions apply when deleting data?
- What restrictions apply when viewing data?
- What rules apply to interfaces?
- What rules apply to conversions?
- What rules apply to reports?
- What rules apply to approvals?
- What rules apply to exceptions?
- What rules can be overridden?
- Who can override them?
- When should overrides be logged?
- What statuses are required?

- What flags are required?
- What codes are required?
- What values are valid?
- What values are invalid?
- What determines when a value changes?
- What determines when a process moves to the next step?

Quick Tip

Business rules should be documented separately from general narrative requirements so they can be reviewed, tested, governed, and changed without confusion.

Common Artifacts

- Business rules catalog
- Decision table
- Validation rules table
- Policy mapping
- Rules traceability matrix
- Approval matrix
- Exception handling model

CHAPTER 20

External Resource, Stakeholder, and Project Team Questions

Projects are affected by people, groups, systems, vendors, regulators, and downstream consumers.

Ask:

- Who is the sponsor?
- Who is funding the project?
- Who owns the business outcome?
- Who owns the process?
- Who owns the system?
- Who owns the data?
- Who are the subject matter experts?
- Who are the users?
- Who currently performs the work?
- Who will perform the work after implementation?
- Who will interact directly with the system?
- Who will interact indirectly with the system?
- Who will send information to the solution?
- Who will receive information from the solution?
- What systems will interact with the solution?
- What vendors are involved?
- What suppliers are involved?
- What distributors are involved?
- What government agencies are involved?
- What regulatory agencies are involved?
- What teams must be consulted?
- What teams must be informed?
- What teams must approve?
- What groups will be impacted downstream?
- Who will support the solution?

- Who will train the users?
- Who will maintain the documentation?
- Who will handle production issues?
- Who will champion the project?

Quick Tip

The result of stakeholder questions should be a comprehensive map of people, systems, teams, organizations, interfaces, and decision authorities involved in the project.

Common Artifacts

- Stakeholder register
- RASCI matrix
- Communication plan
- Interface inventory
- Project team map
- Support model
- Approval matrix

CHAPTER 21

User Interface, Screen, and Report Questions

Screens and reports are often where users experience the requirements most directly.

Ask:

- What data will appear on the screen?
- What actions are allowed on each field?
- Can the user create data?
- Can the user update data?
- Can the user cancel data?
- Can the user delete data?
- Can the user search data?
- Can the user view data?
- Which fields are required?
- Which fields are optional?
- Which fields are conditionally required?
- Which fields are read-only?
- What validation rules apply to each field?
- What labels should be used?
- What buttons are required?
- What menus are required?
- What links are required?
- What help features are required?
- What sections should appear on the screen?
- What headers and footers are required?
- What reports are required?
- What report filters are required?
- What report outputs are required?
- Who can view each report?
- Who can export each report?
- What error messages are required?

- What warning messages are required?
- What confirmation messages are required?
- What response time is expected?
- How should the layout support the workflow?

Quick Tip

Screens should not simply display data. They should support the way users think, decide, and work.

Common Artifacts

- Screen mockups
- Wireframes
- Report specifications
- Field-level requirements
- Error message catalog
- Prototype
- Security matrix
- Usability review notes

CHAPTER 22

Usability Questions

A solution can meet technical requirements and still fail if users cannot use it effectively.

Ask:

- Does the product meet the requirements?
- Does the product meet the objectives?
- Can users accomplish their tasks?
- Is the interface simple?
- Is the interface intuitive?
- Do screens follow a logical flow?
- Do labels use clear business terminology?
- Are buttons understandable?
- Are error messages helpful?
- Does the system prevent avoidable errors?
- Does the system provide feedback?
- Does the system show progress?
- Does the system support novice users?
- Does the system support expert users?
- Does the system support accessibility needs?
- Can users operate the system without a mouse?
- Is the layout consistent?
- Are related fields grouped together?
- Are users able to work efficiently?
- What tasks take too long?
- What tasks are confusing?
- What tasks require too many clicks?
- What information is hard to find?
- What training will be needed?
- How will usability be measured?

Quick Tip

Ask usability questions in a positive, non-accusing way. Instead of asking, "Why would you do it that way?" ask, "Help me understand why it is done that way."

Common Artifacts

- Usability checklist
- User testing notes
- Accessibility checklist
- Prototype feedback
- User journey map
- Training needs assessment

CHAPTER 23

Quality, Performance, and Non-Functional Questions

Non-functional requirements define how well the solution must operate.

Ask:

- What are the performance requirements?
- What response time is expected?
- What transaction volume is expected?
- What user volume is expected?
- What peak volume is expected?
- What growth is expected over time?
- What accuracy is required?
- What availability is required?
- What downtime is acceptable?
- What reliability is required?
- What scalability is required?
- What maintainability is required?
- What supportability is required?
- What portability is required?
- What interoperability is required?
- What accessibility is required?
- What localization is required?
- What legal requirements apply?
- What compliance requirements apply?
- What audit requirements apply?
- What privacy requirements apply?
- What disaster recovery requirements apply?
- What backup requirements apply?
- What archiving requirements apply?
- What purging requirements apply?
- What is the acceptable tolerance for error?

- What is the business impact of an error?
- What is the business impact of downtime?
- What service-level expectations exist?

Quick Tip

Timeframe your questions. Some requirements change based on deadlines, milestones, calendar dates, aging, anniversary dates, year-end processing, or regulatory cycles.

Common Artifacts

- Non-functional requirements catalog
- Performance requirements
- Service-level agreement
- Capacity model
- Disaster recovery requirements
- Audit requirements
- Operational support model

CHAPTER 24
Security Questions

Security requirements protect people, data, systems, and the organization.

Ask:

- What security levels are required?
- What roles are required?
- What permissions are required?
- What authentication is required?
- What authorization is required?
- What password rules are required?
- What audit requirements exist?
- What logging requirements exist?
- What data protection requirements exist?
- What privacy requirements exist?
- What data is confidential?
- What data is sensitive?
- What data is regulated?
- What data must be encrypted?
- What screens require restrictions?
- What fields require restrictions?
- What reports require restrictions?
- What processes require restrictions?
- What interfaces require restrictions?
- Who can create records?
- Who can update records?
- Who can delete records?
- Who can cancel records?
- Who can purge records?
- Who can view records?
- Who can approve transactions?
- Who can override rules?

- What access should be removed when users change roles?
- What access should be removed when users leave the organization?
- What firewall requirements exist?
- What intrusion detection requirements exist?
- What virus protection requirements exist?
- What regulatory security standards apply?
- What company security policies apply?

Quick Tip

Security should not be added at the end. Security requirements should be discovered early, traced clearly, tested thoroughly, and governed continuously.

Common Artifacts

- Security matrix
- Role-based access model
- Audit requirements
- Privacy requirements
- Compliance mapping
- Data classification model
- Access control requirements

CHAPTER 25

Technical and Solution Questions

Technical requirements define the solution environment, constraints, and implementation boundaries.

Ask:

- What are the physical limitations?
- What are the technology limitations?
- What are the budget limitations?
- Are we restricted to current technology?
- Are specific hardware platforms required?
- Are specific software tools required?
- Are specific programming languages required?
- Are specific database tools required?
- Are specific integration tools required?
- Are specific cloud platforms required?
- Are there network requirements?
- Are there connectivity requirements?
- Are there browser requirements?
- Are there mobile requirements?
- Are there platform requirements?
- Are there hosting requirements?
- Are there performance constraints?
- Are there architecture constraints?
- Are there vendor constraints?
- Are there licensing constraints?
- Are there support constraints?
- Are there deployment constraints?
- Are there configuration requirements?
- Are there environment requirements?
- Are there monitoring requirements?
- Are there operational requirements?

Quick Tip

Technical questions should not dominate discovery too early. First understand the business problem, process, data, rules, users, and outcomes. Then evaluate the technical solution.

Common Artifacts

- Technical requirements
- Architecture constraints
- Environment requirements
- Integration requirements
- Deployment requirements
- Infrastructure requirements
- Configuration requirements

CHAPTER 26

Gap Analysis Questions

Gap analysis compares the current state to the desired future state.

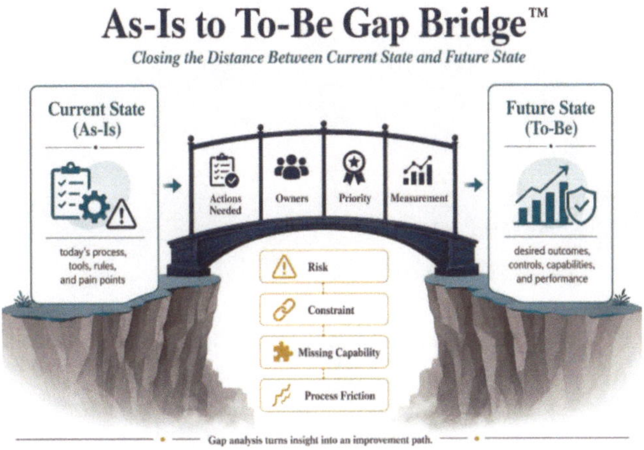

Ask:

- Where are we now?
- Where do we want to be?
- What will happen if we do nothing?
- What currently works well?
- What should not change?
- What are the weaknesses?
- What are the opportunities?
- What are the threats?
- What policies exist today?
- What procedures exist today?
- What standards exist today?
- What tools exist today?
- What documentation exists today?
- What training exists today?
- What skills exist today?
- What skills are missing?

- What systems exist today?
- What systems are missing?
- What data exists today?
- What data is missing?
- What controls exist today?
- What controls are missing?
- How do we close the gap?
- What must we buy?
- What must we build?
- What must we configure?
- What must we retire?
- What must we improve?
- What can be reused?
- What must be replaced?
- How will we measure gap closure?
- How will we know the gap has been reduced?

Quick Tip

Gap analysis should produce action. If the gap is identified but not assigned, prioritized, funded, or tracked, it remains only an observation.

Common Artifacts

- Gap analysis worksheet
- Current-to-future comparison
- Improvement roadmap
- Action plan
- Risk log
- Process improvement backlog
- Capability assessment

PART IV

REQUIREMENTS INTELLIGENCE IN PRACTICE

CHAPTER 27

Sunny Day, Rainy Day, and Exception Path Thinking

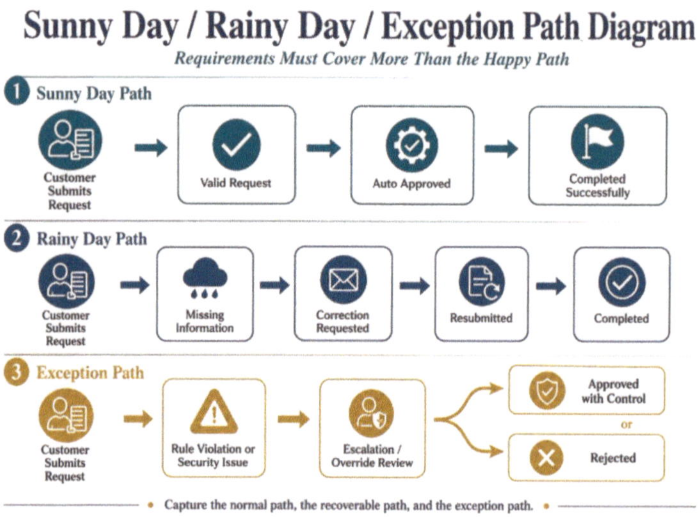

Requirements are incomplete when they describe only the ideal scenario.

Every process has at least three views:

- The sunny day path
- The rainy day path
- The exception path

The **sunny day path** describes what happens when everything works as expected.

The **rainy day path** describes what happens when common problems occur.

The **exception path** describes what happens when something unusual, risky, invalid, delayed, unauthorized, or unexpected occurs.

For each requirement, ask:

What happens when everything works?
What happens when required information is missing?
What happens when information is incorrect?
What happens when the user lacks permission?
What happens when the system is unavailable?
What happens when another system does not respond?
What happens when the transaction is rejected?
What happens when approval is delayed?
What happens when the customer changes their mind?
What happens when a deadline passes?
What happens when a rule is overridden?
Quality requirements include all three paths.

CHAPTER 28

Requirements Buckets

As answers are gathered, they should be placed into requirements buckets.

Requirements buckets help the analyst organize information before it becomes overwhelming.

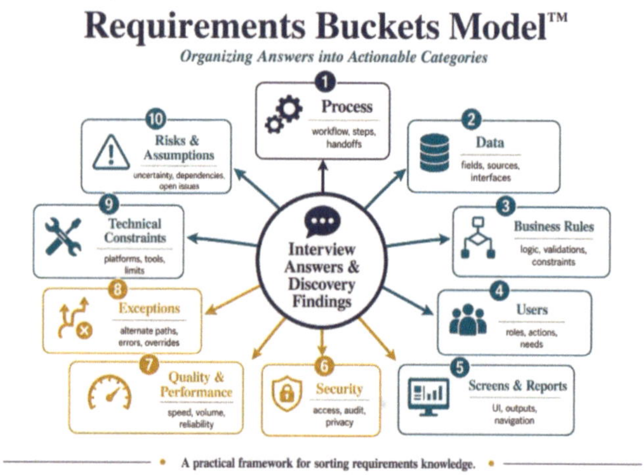

Common buckets include:

Business objectives

Scope

Stakeholders

Current-state process

Future-state process

Features

Functions

User roles

Business rules

Data

Interfaces

Reports

Security

Usability

Quality

Performance

Exceptions

Technical constraints

Risks

Assumptions

Dependencies

Open questions

Out-of-scope items

Future enhancements

This structure prevents requirements from becoming a pile of disconnected notes.

The analyst should continuously ask:

- What type of answer did I just receive?
- Where does this information belong?
- What artifact should be updated?
- Who needs to validate this?
- What requirement does this create?
- What test case will prove it works?

CHAPTER 29

Translating Questions into User Stories, Use Cases, and Requirements Tables

The same answer can be expressed in different formats depending on the project method.

A stakeholder may say:

"Our supervisors need to approve exceptions before they are processed."

That answer may become:

A business rule:

Supervisors must approve all exception transactions before processing.

A user story:

As a supervisor, I need to review and approve exception transactions so that only authorized exceptions are processed.

A use case step:

The system displays the exception transaction to the supervisor. The supervisor approves or rejects the transaction.

An acceptance criterion:

Given an exception transaction exists, when a supervisor approves it, then the system records the approval, date, time, and supervisor ID before allowing processing to continue.

A test case:

Verify that exception transactions cannot proceed without supervisor approval.

A security requirement:

Only users assigned the Supervisor role may approve exception transactions.

A reporting requirement:

The system shall provide an exception approval report showing transaction ID, approval status, supervisor ID, approval date, and rejection reason where applicable.

Questions produce answers.

Answers produce requirements.

Requirements produce design, development, testing, and operational controls.

CHAPTER 30

From Question to Requirement to Test Case Traceability Chain™

Traceability protects the project from losing meaning.

Every important requirement should be traceable to:

The original question

The stakeholder who answered

The business objective

The process step

The data element

The business rule

The design component

The test case

The acceptance criteria

The approval decision

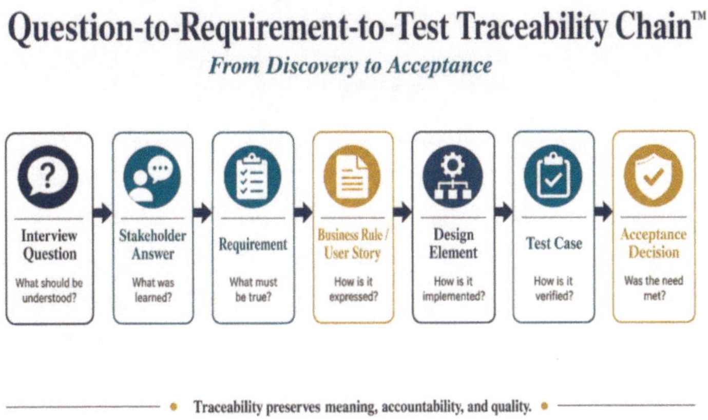

Traceability answers the question:

Why are we building this?

Without traceability, requirements become vulnerable to opinion, memory, politics, and assumption.

With traceability, the team can determine:

- Who requested the requirement
- Why it matters
- What it supports
- How it will be tested
- Whether it is still needed
- What will be affected if it changes

Traceability is not bureaucracy.

Traceability is memory.

CHAPTER 31

Requirements Risk and the Cost of Ambiguity

Ambiguity is one of the greatest risks in requirements work.

Words such as fast, simple, easy, flexible, secure, intuitive, user-friendly, real-time, automated, seamless, scalable, robust, and efficient may sound useful, but they are incomplete unless defined.

Ask:

- What does fast mean?
- What does simple mean?
- What does secure mean?
- What does real-time mean?
- What does automated mean?
- What does scalable mean?
- What does user-friendly mean?
- What does efficient mean?
- How will we measure it?
- Who decides whether the requirement has been met?

A vague requirement cannot be reliably built.

A vague requirement cannot be reliably tested.

A vague requirement cannot be reliably accepted.

The analyst must convert vague language into measurable meaning.

CHAPTER 32

Requirements Review and Sign-Off

Requirements review is the point where understanding becomes agreement.

Before sign-off, confirm:

The business problem is clear.

The project scope is clear.

The stakeholders are identified.

The current state is understood.

The future state is defined.

The business drivers are documented.

The users are identified.

The features are prioritized.

The business rules are documented.

The data requirements are documented.

The process requirements are documented.

The exception paths are documented.

The security requirements are documented.

The quality requirements are documented.

The technical constraints are documented.

The risks are documented.

The assumptions are documented.

The open issues are documented.

The out-of-scope items are documented.

The requirements are traceable.

The acceptance criteria are clear.

The appropriate stakeholders have reviewed and approved.

Sign-off does not mean requirements will never change.

It means the team has reached a shared understanding at a point in time.

PART V

INTELLIGENT SYSTEMS PRESS FRAMEWORK CONNECTION

CHAPTER 33

Requirements Analysis and Intelligent Process Improvement

Intelligent Process Improvement begins with understanding how work actually happens.

Requirements analysis contributes to process improvement by identifying:

Current process steps

Pain points

Manual workarounds

Bottlenecks

Redundant activities

Rework loops

Unclear ownership

Broken handoffs

Missing controls

Outdated rules

Unsupported decisions

Data problems

User frustrations

Policy gaps

Automation opportunities

A requirements analyst is often the first person to see the full picture across departments, systems, users, and decision points.

That makes requirements analysis a process improvement discipline.

The analyst does not merely ask what the system should do.

The analyst asks how the organization should work.

CHAPTER 34

Requirements Analysis and Process Intelligence

Process Intelligence requires visibility into how work flows across people, systems, data, rules, and outcomes.

Requirements questions help create that visibility.

When a team asks:

Who performs the work?

What data is used?

What rule applies?

What exception occurs?

What approval is needed?

What system is involved?

What outcome is expected?

What metric proves success?

The team is not only defining requirements.

It is building process intelligence.

Each answer becomes part of the organization's operational knowledge.

When captured correctly, requirements become more than project documentation. They become reusable intelligence for future improvement, automation, governance, and decision-making.

CHAPTER 35

Requirements Analysis and the 12-Step IPI Model

The 12-Step Intelligent Process Improvement model depends on disciplined discovery.

Requirements questions support that model by helping teams:

Define the process

Identify the problem

Clarify ownership

Map the current state

Define the future state

Identify gaps

Document rules

Measure performance

Validate improvements

Control change

Maintain alignment

Prevent process drift

Without requirements discovery, process improvement becomes guesswork.

With structured questioning, process improvement becomes evidence-based.

CHAPTER 36

Requirements Analysis and Legacy Structured Software Development

Structured software development placed heavy emphasis on requirements, documentation, analysis, design, testing, review, and control.

This book preserves that discipline.

Even in Agile, DevOps, AI-enabled, and cloud-based environments, the need for clarity has not disappeared.

Teams may deliver faster, but they still need to know:

What problem are we solving?

Who is the user?

What is the process?

What is the rule?

What is the data?

What is the exception?

What is the risk?

What is the acceptance criterion?

How will we know this works?

Modern delivery does not eliminate requirements.

It increases the need for better questions.

CHAPTER 37

Requirements Analysis for Agile, DevOps, AI, and Enterprise Governance

Requirements analysis must evolve without losing its foundation.

In Agile environments, questions help refine user stories, acceptance criteria, backlog priorities, dependencies, and release goals.

In DevOps environments, questions help define deployment, monitoring, reliability, operational support, security, and recovery requirements.

In AI-enabled environments, questions help define data quality, model purpose, decision boundaries, human oversight, explainability, bias risk, auditability, and governance.

In enterprise governance environments, questions help define compliance, accountability, traceability, risk, policy alignment, and executive visibility.

The question remains the common thread.

Whether the organization is using waterfall, Agile, DevOps, AI, or hybrid delivery, the discipline of asking better questions remains essential.

APPENDIX A

Master Requirements Question Checklist

Use this checklist to guide requirements discovery.

Current State

What exists today?

What works well?

What does not work well?

Who performs the work?

What systems are used?

What data is captured?

What rules apply?

What problems occur?

What risks exist?

Future State

What should change?

What should remain?

What should be automated?

What should be improved?

What outcomes are expected?

What capabilities are required?

What success factors matter?

Scope

What is included?

What is excluded?

What assumptions exist?

What constraints exist?

What dependencies exist?

What decisions have been made?

What risks threaten success?

Business Drivers

Why is this project needed?

What value will it create?

What cost will it reduce?

What risk will it reduce?

What revenue will it support?

What compliance need will it satisfy?

What happens if nothing changes?

Users and Features

Who will use the solution?

What will they do?

What roles exist?

What permissions exist?

What features are required?

What functions are required?

What tasks must be supported?

Exceptions

What happens when the normal path fails?

What errors can occur?

What alternate paths exist?

What overrides are allowed?

What must be logged?

What must be escalated?

Data

What data is needed?

Where does it come from?

Who owns it?

What is mandatory?

What is optional?

What is validated?

What is retained?

What is converted?

What is reported?

Workflow

What starts the process?

What ends the process?

What steps occur?

Who performs each step?

What handoffs occur?

What approvals occur?

What delays occur?

What outputs are created?

Business Rules

What decisions are made?

What rules apply?

What calculations are required?

What validations are required?

What restrictions exist?

What policies apply?

What exceptions exist?

Stakeholders

Who sponsors the work?

Who owns the process?

Who owns the data?

Who owns the system?

Who uses the solution?

Who supports it?

Who approves it?

Who is impacted?

User Interface

What screens are needed?

What fields are needed?

What buttons are needed?

What reports are needed?

What messages are needed?

What layout supports the workflow?

Usability

Can users complete their tasks?

Is the flow logical?

Are labels clear?

Are errors helpful?

Is accessibility addressed?

Is training needed?

Quality and Performance

How fast must it be?

How available must it be?

How reliable must it be?

How scalable must it be?

How accurate must it be?

What volumes are expected?

What downtime is acceptable?

Security

Who can access what?

Who can create records?

Who can update records?

Who can delete records?

Who can approve records?

What must be encrypted?

What must be audited?

What policies apply?

Technical Solution

What platforms are required?

What tools are required?

What integrations are required?

What environments are required?

What limitations exist?

What architecture constraints exist?

Gap Analysis

Where are we now?

Where do we want to be?

What is missing?

What must change?

What can be reused?

What must be replaced?

How will we close the gap?

APPENDIX B

Interview Planning Template

Project Name:
Interview Date:
Facilitator:
Stakeholder Name:
Stakeholder Role:
Department:
Interview Purpose:

Topics to Cover

Current state
Future state
Pain points
Scope
Users
Data
Workflow
Business rules
Exceptions
Security
Reports
Quality
Risks
Open questions

Questions to Ask

1.
2.
3.

4.

5.

Documents to Request

Process maps

Reports

Forms

Screenshots

Spreadsheets

Policies

Procedures

Training materials

System documentation

Audit findings

Help desk tickets

Follow-Up Items

Item:

Owner:

Due Date:

Status:

APPENDIX C

Stakeholder Discovery Template

Use this redesigned worksheet to capture stakeholder role, responsibility, decision authority, and major concerns without crowding the page.

Stakeholder	Role / Department	RASCI Role	Contact / Team	Key Concern	Notes / Decision Authority

Project Name:

Key Questions:

Who owns the process?

Who owns the data?

Who owns the system?

Who funds the project?

Who approves the requirements?

Who uses the solution?

Who supports the solution?

Who is impacted downstream?

APPENDIX D

Requirements Traceability Matrix Template

Use this compact matrix to preserve traceability from the original question through business value, design, test coverage, and status.

Req ID	Requirement Description	Source Question / Stakeholder	Business Objective / Priority	Design Reference / Test Case	Status
REQ-001					
REQ-002					
REQ-003					
REQ-004					

APPENDIX E

Business Rules Table Template

Use this worksheet to separate decision logic, triggers, actions, exceptions, ownership, and verification.

Rule ID	Business Rule	Condition / Trigger	Action / Decision	Exception / Owner	Source / Test Case
BR-001					
BR-002					
BR-003					
BR-004					

APPENDIX F

Gap Analysis Worksheet

Use this worksheet to compare current state and future state, identify the gap, assign priority, and define ownership for closure.

Area	Current State	Future State	Gap / Risk	Priority / Impact	Action Needed / Owner
Process					
Data					
Rules					
Systems					
People					
Security					

APPENDIX G

Question-to-Artifact Mapping Matrix

How Better Questions Produce Better Deliverables

#	Question Category	What It Reveals	Primary Artifact
1	As-Is	current-state reality	workflow diagram
2	To-Be	future-state intent	future-state model
3	Scope	boundaries and assumptions	scope statement
4	Business Drivers	value and priority	business case
5	User Requirements	roles, needs, features	use cases or user stories
6	Exceptions	alternate paths and edge cases	exception log
7	Data	information needs and sources	data dictionary
8	Business Rules	decision logic and constraints	rules catalog
9	Security	access and control	security matrix
10	Gap Analysis	current-to-future gaps	gap analysis worksheet

Map every important question to a usable project artifact.

APPENDIX H

Glossary of Requirements and Process Engineering Terms

Acceptance Criteria

Conditions that must be satisfied for a requirement, user story, feature, or function to be accepted.

Actor

A person, system, organization, or role that interacts with a process or solution.

As-Is Process

The current-state process as it operates today.

Business Analyst

A professional who investigates business needs, elicits requirements, analyzes processes, documents requirements, and supports solution delivery.

Business Rule

A policy, condition, calculation, restriction, or decision logic that governs behavior, data, or process flow.

Current State

The existing environment, process, system, or workflow before change.

Data Dictionary

A structured description of data elements, including definitions, formats, values, rules, and ownership.

Exception Path

A scenario that occurs when the normal flow cannot proceed as expected.

Future State

The desired process, system, workflow, or operational condition after change.

Gap Analysis

A comparison between current state and future state to identify what must change.

Happy Path

The normal or ideal process flow where everything works as expected.

Non-Functional Requirement

A requirement that defines how a system performs or operates rather than what function it performs.

Process Intelligence

An enterprise operating capability that uses process visibility, data, rules, metrics, governance, and feedback loops to align execution with business outcomes.

Requirement

A documented need, condition, capability, rule, or constraint that a solution must satisfy.

Requirements Traceability

The ability to connect a requirement to its source, objective, design, test case, and approval.

Scope

The defined boundaries of what is included and excluded from a project or solution.

Stakeholder

Any person, group, system, or organization that affects or is affected by the project or solution.

To-Be Process

The desired future-state process.

Use Case

A description of how an actor interacts with a system to accomplish a goal.

User Story

A concise statement of user need, commonly written from the perspective of a role or user.

APPENDIX I

Intelligent Systems Press Series Roadmap

The Intelligent Systems Press process engineering series is designed to guide readers from requirements discovery to enterprise intelligence.

1. Project Requirements Analysis Book of Questions

The requirements discovery foundation.

Focus: asking better questions, defining clearer requirements, and understanding the problem before designing the solution.

2. Feng Shui for Intelligent Process Improvement

The process balance foundation.

Focus: restoring structure, order, flow, and discipline to process environments.

3. 12 Steps to Intelligent Process Improvement

The improvement method.

Focus: defining, fixing, measuring, maintaining, and governing process improvement.

4. Process Intelligence

The enterprise operating layer.

Focus: connecting Agile, DevOps, AI, governance, and executive strategy through intelligent process alignment.

5. Process Intelligence Playbook

The stakeholder execution guide.

Focus: helping executives, engineers, product owners, project managers, Scrum Masters, QA teams, and process leaders apply the framework in real-world environments.

Closing Statement

The quality of a project is often determined before design begins, before development starts, before testing finds defects, and before users ever see the final solution.

It is determined when the first questions are asked.

A project team that asks shallow questions builds from assumption.

A project team that asks disciplined questions builds from understanding.

A project team that listens, probes, validates, traces, and improves creates more than requirements.

It creates intelligence.

And intelligence is the beginning of better process, better delivery, better governance, and better business outcomes.

Questions are not the interruption. Questions are the answer.

www.ingramcontent.com/pod-product-compliance
Lightning Source LLC
LaVergne TN
LVHW051129210926
848482LV00001B/30